Exploring the Night Sky with Binoculars

A Companion to *The Night Sky* Star Dial
by David Chandler

P.O. Box 309, La Verne, CA 91750

Paintings by Don Davis

Maps and Line Drawings by David Chandler

ISBN 0-9613207-0-2

Table of Contents

Appendices

Getting in Touch with the Sky

Nature extends beyond the realm of trees and wildlife. In its grandest scale the natural universe can be seen arching overhead on clear, dark nights. Exploring the night sky is a fundamental experience that should be shared by us all, as much as the experience of strolling through the woods or witnessing a rainbow or thunderstorm. The sky above can put us in touch with our roots in the cosmos and set our daily concerns in a broader context. It poses the ultimate questions of who we are and what our place is in the scheme of things. To watch the sky roll overhead requires that we temporarily set aside our daily routine and pace ourselves with the universe.

The spectacle of the night sky is not reserved for the astronomer. The sky on a dark, clear night is impressive even to the unaided eye. With the addition of a simple pair of binoculars you can explore the heavens to a far greater extent than you may realize. In fact the same binoculars you may use for watching birds or football games can provide a better introduction to the sky than most telescopes sold in department stores advertizing high power and costing several hundred dollars. A good telescope is a wonderful resource, and guidelines for selecting telescopes will be provided later, but your observing experience will be far richer if you start with binoculars.

Why Binoculars?

Binoculars have several important advantages over the typical 'astronomical' telescope, particularly for the beginner.

Binoculars use low power. This may not seem to be an advantage, but in many ways it really is. Low power means a wider field of view and a brighter image. Many of the interesting objects in the sky are not small, they are just faint. The Andromeda Galaxy, for instance, is about five times as large in the sky as the full moon! It is not commonly noticed only because it is faint. Binoculars show it beautifully. Excess magnification cuts down the field of view, and by spreading the light over a larger area, makes the image fainter. Certain star clusters and nebulae, comets, star clouds and dust lanes in the Milky Way may actually be seen better in binoculars than in any other optical instrument. A typical telescope has a field of view at low power of about one degree. That is like looking at the world through a soda straw! The sky covers over 40,000 square degrees, so if you are going to use a telescope with a one degree field of view, you have to have a pretty good idea ahead of time where to look and how to get the telescope aimed where you want it. Randomly sweeping the sky with a telescope usually produces little of interest and can lead to discouragement for a new telescope owner. A typical binocular field of view, on the other hand, is about seven degrees across and covers about fifty times as much area as a one degree telescopic field. Searching for an object is much easier, and leisurely sweeping large portions of the sky is a richly rewarding option. With binoculars you can experience the joy of discovery from the start.

Another problem with astronomical telescopes for the beginner is that they seem 'magical'. When a large telescope is open for viewing by the public it is typical for someone to ask to see the moon on nights when, as anyone can see, the moon is not even up. The view through a large telescope is so different from our usual experience of the sky that one can feel as though he were entering a magical realm, and not fully grasp what is being seen. Binoculars provide a softer transition. You understand what you see in binoculars because binoculars are common household items used in very ordinary circumstances. When binoculars are turned to the night sky there is continuity of experience and hence better understanding. After seeing a galaxy or nebula in binoculars the transition to a telescope comes more naturally.

Finally, binoculars are the ideal 'first telescope' because they are readily available. Even if you don't already own a pair, good quality binoculars are easy to find and are moderately priced due to mass production. A good pair of binoculars, rather than a telescope, should be your first astronomical investment.

Choosing and Using Binoculars

Binocular Selection

Not all binoculars are created equal. Here is some background information, whether you plan to buy binoculars or become more familiar with ones you already own.

Binoculars are rated with a pair of numbers such as 6x30, 7x35, 7x50, 10x50, etc. The first number is the magnification or power, and the second number is the size of the front lenses in millimeters. Both numbers are important. The larger the lenses, the more light will be gathered, so the bigger the better. Increased magnification can be a good thing, but greater magnification requires correspondingly larger lenses to avoid dimming the image. Also, if the magnification is too high, it will be harder to hold the image steady. Weight is also a consideration. At some point a tripod becomes highly desirable. 7x50 binoculars are generally considered the best general purpose binoculars for hand-held nighttime use. 10x70, 11x80, and 20x80 are also available and excellent for astronomy, but they are not easy to use hand-held because of their weight.

When buying binoculars, especially inexpensive or second hand ones, be sure to check out the optical system. If the lenses are of good quality and aligned properly you should be able to get sharp images and view for extended periods without eyestrain. Binoculars are subject to alignment problems, especially if they have been dropped or otherwise abused. Check the optical system by looking backward through each barrel. You should be able to see straight through with every optical element centered. Loose lenses, chipped glass, or gross misalignment will usually be detectable in this way. Problems could also be caused by misalignment of one barrel relative to the other. Probably the most essential test is viewing comfort. When buying binoculars look through them, preferably for five minutes or more, focusing on objects at various distances. Don't buy them if they cause eye strain. The acid test of optical quality is the appearance of star images. Stars should focus to sharp, clean points of light.

Proper Use

Focus is a prime consideration. Some binoculars have independent focus adjustments for each eye. If this is the case simply focus one eye at a time. The only inconvenience might be if the binoculars are going to be passed around and refocused repeatedly for different people. In that case you might want to make a mark for the proper setting for your eyes.

Most binoculars have a center focus knob with a separate adjustment for one eye only. If yours are of this type, use the center knob to focus the non-adjustable barrel, then adjust the single eye adaptor until both eyes are in focus.

A good focus is a relaxed focus. The skill is to learn to

relax your eyes first and let the binoculars accomodate to them. This is not as easy as it sounds. Your eyes will try to adapt to any image presented to them. When you intentionally relax your eyes, letting the image blur if it will, it causes a mild 'spaced out' or dreamy sensation. Notice which way the eyepieces move as you turn the knob. A relaxed focus is easiest to obtain if you start with the eyepieces too far out and slowly bring them in until the stars look sharp. Play around, taking your binoculars in and out of focus, and learn the feeling of a relaxed focus.

If you wear glasses you will be more comfortable without them unless you have severe astigmatism. Binoculars can correct for near or far sightedness, but not astigmatism. Contact lenses present no problem.

The next skill to learn is pointing. This sounds childishly simple, but when looking at the stars one field of view will look like any other at first. The problem is compounded because you will suddenly see many more stars than before and the magnification may be deceptive. For instance, in looking at the Big Dipper you will generally see only one of the major stars in the field at a time! A good exercise is to practice tracing out familiar constellations with the binoculars. One way to locate an object is to aim at the horizon directly below it and sweep straight up until it comes into view. Another way is to look rigidly at the desired spot in the sky without altering your gaze as you set the binoculars in place before your eyes. There is usually a tendency to aim too low. If you are trying to find a faint object, aim at a nearby bright star and 'star hop' to your goal. It takes a little practice but soon you will become proficient.

Experienced amateur astronomers come to realize that seeing is itself a skill to be acquired. Your eyes are primarily adapted for seeing in bright light. The center of vision on the retina of your eyes is densely packed with color detectors called cones. The surrounding areas of the retina contain primarily black and white light detectors called rods. Rods are very sensitive to faint light, but cones are not. That is why at night everything appears in shades of grey. To see a very faint object it is often necessary to use averted vision, looking slightly to one side so the light falls on the more sensitive rods. The skill comes in learning to focus your attention on something without focusing your eyes on it. It is hard at first to feel you are really seeing something when you are not looking directly at it.

Finally, remember you are going to be standing around at night in the cold. You may be in Southern California, but dress for the arctic. You need covering for your head, warm socks, heavy coat, and perhaps long underwear, and gloves. Staying comfortable is essential to a leisurely time communing with the sky. Oh yes, don't forget a thermos of coffee, plenty of munchies, lawn chairs, and some friendly companionship! Make an event out of it. Find opportunities to drive into the country on moonless nights, far from lights of any kind, for the best viewing conditions you can find. Watching the sky roll majestically overhead can be an incomparable experience.

Where to Start

Take your binoculars out at night and see what you can see. Look at the brightest stars. They will appear simply as points of light, unless you accidentally happen upon a planet. Planets are relatively nearby and can appear as disks, but stars are so far away they appear as points even in the largest telescopes. As you scan around the sky you will notice that many more stars are visible with binoculars than with the unaided eye. This will be particularly true in the region of the Milky Way. If you are under a fairly dark sky with no moon or city lights you may also find a number of fuzzy or clumpy things which we shall discuss later.

Randomly sweeping the sky you should be able to find quite a few interesting star clouds and clumps, but to talk about specific objects you need some way to find your way around and know where you are looking. The simplest method, for our purposes, is for you to become familiar with the constellations. These star patterns allow you to recognize areas of the sky much as you recognize countries on a map of the earth.

Learning the Constellations

Learning the constellations is not as difficult as some people imagine. Here are some pointers.

■ First, use a good constellation chart. This booklet was designed as a companion to *The Night Sky*, a star dial

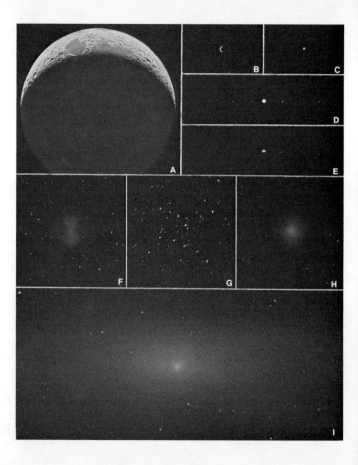

Scale drawings of various objects as seen in binoculars. For correct scale in 10x binoculars, hold page at 18 inches. For 7x binoculars, hold page at 26 inches.
A. Crescent moon, B. Crescent phase of Venus, C. Mars, D. Jupiter with its moons, E. Saturn with its moon Titan, F. The Dumbbell Nebula, G. Typical open cluster, H. Typical globular cluster, I. The Andromeda Galaxy.

designed by the author. A star dial is convenient because it can match the rotation of the sky for any night of the year and any hour of the night. An alternate approach is to use a different chart for each month. *Sky and Telescope Magazine* has an excellent monthly constellation chart in each issue. See the Resource Guide in Appendix C for addresses in either case.

■ To read star charts astronomers usually cover their flashlights with red acetate (available in art stores) or even simply a brown paper bag to keep from exposing their eyes to bright light. Your eyes take a long time to adapt completely to the dark, and excess light will hinder the process.

■ Learn the first magnitude stars (ie. the brightest stars) first. They stand out plainly and there are only a few of them. You should note that the planets Mercury, Venus, Mars, Jupiter, and Saturn are usually as bright or brighter than first magnitude stars. Once you know the first magnitude stars you can quickly pick out the planets, which look like bright stars out of place.

■ The constellations come next. Lay aside any expectation that the constellations should look like their name-sakes. The stars form random patterns on the sky. There are no mythological beasts. The point is not to make the stars fit into preconceived patterns, but to remember the patterns of stars as they actually appear in the sky. Pictures or abstract patterns, however fanciful, are useful as aids to the memory, but only if they are not taken too seriously. Identify a star grouping then ask yourself what it looks like to you.

■ Learn the constellations in the context of their neighbors, not from their shapes alone. There are many places in the sky where stars can be formed into 'little dippers'. Constellation names refer to specific groups of stars. Learn them as you would countries on a map of the earth.

■ Look up at the sky every night. You don't have to live in the country. In fact, with city lights there will be fewer faint stars to confuse you. Learn the first magnitude stars the first night and figure out if any planets are visible. Retrace your steps each night and add a few constellations, using the first magnitude stars as jumping off points. Within perhaps a week you should be able to find any constellation in the sky with the help of a star chart. Soon they will come automatically. Binoculars can help you trace out the fainter stars in the constellations if you live in the city. Since only half of the sky is visible at any one time you will have to add constellations as the seasons progress. You will thus become familiar not only with the patterns, but the rotation of the sky as well.

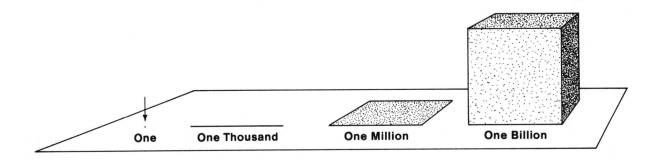

One   One Thousand   One Million   One Billion

A Brief Cosmic Geography Lesson

You will see some impressive sights when you start exploring the sky with binoculars. Some of the immediate questions that arise are phrased in the form, 'How many . . .?', 'How big . . .?', 'How far . . .?', and 'How old . . .?'. The difficulty comes in knowing how to understand the answers once they are given. How do you visualize a million or a billion of anything? To comprehend numbers such as these you need a set of mental pictures. If your experience with numbers is so negative that you are getting an anxiety attack already, you might want to skip to the next section. If you would like to come to grips with the numbers of astronomy, the following mental pictures may help.

Big Numbers

Most people can visualize a hundred of something. You can count to 100, and you think in terms of dollars and cents all the time. To visualize a thousand of something, compare a grain of sand to a yard stick. A thousand course sand grains lined up would measure about a yard. This picture depends, of course, on how big each grain of sand is, but approximate numbers are good enough for our purposes. The main idea is to get in the right ballpark.

A million is a thousand thousands, so picture a thousand rows of a thousand grains of sand. Side by side these rows would cover an area of one square yard. Swept up, this much sand would just about fill a quart container. Stretched out into a line it would extend a little over half a mile.

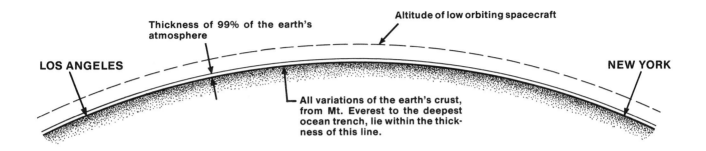

Thickness of 99% of the earth's atmosphere

Altitude of low orbiting spacecraft

LOS ANGELES

NEW YORK

All variations of the earth's crust, from Mt. Everest to the deepest ocean trench, lie within the thickness of this line.

A billion grains of sand would be a thousand layers of a million grains each, and would occupy a cubic yard. A trillion grains of sand (1000 cubic yards) would fill a container 10 yards on each side. We won't talk about any numbers larger than this.

Now let's apply our number images to the cosmos. Our galaxy has somewhere between 100 billion and a trillion stars. Think of something that has a volume of 100 to 1000 cubic yards, a large swimming pool perhaps, and imagine it filled with sand. That's a lot of sand! It is an interesting coincidence that the observable universe has about as many galaxies as a single galaxy has stars, so we can use the same picture twice.

Sizes and Distances

Let's work our way up through the cosmic distance scale. The earth is about 8000 miles across. The moon is about a fourth as big and a quarter million miles away. Jupiter, the largest planet, is about ten times the size of the earth, and the sun is about ten times the size of Jupiter. That makes the sun just under a million miles across. The earth is nearly 100 million miles from the sun, so if we pictured the sun to be the size of our standard grain of sand, the earth would be about 3½ inches away. The earth would be almost too small to see and the entire orbit of the moon would be half the size of the sand grain representing the sun. Jupiter would be 1½ feet from the sun, and Pluto would be about 12 feet away. Thus the whole solar system, as we

Earth

Moon

←———————————— 240,000 miles ————————————→

have pictured it, would easily fit on half a tennis court.

The solar system is mostly just a star, the sun. The planets, their moons, and a few smaller chunks of rock and ice, called asteroids and comets, respectively, are some of the debris left over from the sun's formation. One of those little specks is pretty important to those of us who call it home, but in terms of comparative size, it is a speck nonetheless.

The next star, on this scale, would be another sand grain 15 miles away. This is typical of the spacing between stars in our part of the galaxy. The galaxy as a whole, as we said earlier, contains between a hundred billion and a trillion stars. Now picture spreading out all that sand into a disk with 15 miles between grains. The model itself would

extend out beyond the moon! To keep going beyond this point we need to change the scale of our model. Picture our galaxy reduced now to be the size of a nickel. On this revised scale the Andromeda galaxy, our nearest large neighbor, would be the size of a quarter about 18 inches away. The entire observable universe would extend outward nearly three miles in all directions. What lies beyond that is hard to say.

Time

Space and time are inextricably woven together. As we look out in space the light we are seeing took time to get here, so we are seeing light that originated in the past. Thus as we look out in space we are also looking back in time.

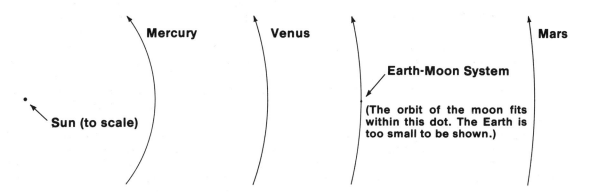

Mercury

Venus

Mars

Earth-Moon System

(The orbit of the moon fits within this dot. The Earth is too small to be shown.)

Sun (to scale)

Light, traveling at 186,000 miles per second takes a little over 1 second to come from the moon, 8 minutes to come from the sun, 5 hours to come from Pluto, 4½ years to come from the next star, 100,000 years to cross the galaxy, and 2 million years to come from the Andromeda galaxy. The universe is estimated to be be about 20 billion years old, so however large the universe may be, we cannot observe out to distances greater than 20 billion light years.

Picture a time scale where the size of a grain of sand represents a million years. Then a yard represents a billion years and 20 yards represents the estimated age of the universe. The sun and its planets formed about 4.6 billion years ago, not at the beginning of time as many people assume. The sun has an estimated stable life span of 10 or 12 billion years, so we are nearly half way through the sun's life cycle. The sun revolves around the galaxy in a couple hundred million years, so we have been around about twenty times since the sun and planets formed. The most massive stars are like cosmic flash bulbs. They burn their fuel at a tremendous rate and have life spans of only about a million years, which carries them only a few degrees around the galaxy before they burn out.

'How do we know all of this?', you ask. If you are interested in digging into the meat of astronomy there are many well written, non-mathematical astronomy texts available today. See the Resource Guide in Appendix C for suggestions.

Now, let's get back to observing.

The Sun's Family

Binoculars can probe deep into space, but there is a lot to see even in our immediate vicintity. Our closest neighbor, the moon, is the easiest target, and a very rewarding one at that. The planets come next. Galileo was the first person to turn a telescope toward these heavenly bodies, and he made an impressive list of discoveries. What he discovered you can discover for yourself, realizing that your binoculars probably outperform Galileo's tiny telescopes.

You can go beyond Galileo. The planets Uranus and Neptune were not discovered until after his time, but both are within your range. Only Pluto remains in the exclusive domain of large telescopes. Among the smaller members of the solar system are the asteroids and comets. Several of the asteroids can be tracked with binoculars, and comets are often seen better in binoculars than in any other optical instrument.

The Moon

How closely have you observed the moon? We all recognize the cycle of the moon's phases, but have you observed closely enough to say where and when the moon will be in the sky when it is a thin crescent, or which direction the curved bright edge faces, or what time the full moon rises? Had you noticed that the moon moves its own diameter against the background of stars each hour? You may know that the moon keeps the same face toward the earth at all times, but had you noticed that it wobbles? Pick a feature near the edge of the moon and watch its distance from the edge vary over a week or two.

The most prominent surface features on the moon are the large, dark patches. Galileo called them 'maria', meaning 'seas', but today we know there is no water on the moon. The maria are lava flows that have filled much of the lunar lowlands.

Numerous craters can be seen with binoculars. The best place to see craters is along the terminator, the rough line that separates the lunar day and night. Near the terminator the sunlight comes in at a grazing angle and casts long shadows, highlighting the shape of the terrain. Each night new sections of the lunar surface are featured as the terminator sweeps across the moon. Notice how rough the light colored lunar highlands are compared to the dark lava plains. It is unreasonable to assume that meteorites hit only the highlands, so we must conclude that the lava came more recently than the heaviest meteor bombardments and flooded many of the lowland craters. Can you find ghost rings of flooded craters? As the terminator sweeps across the maria look also for ripples in the lava called wrinkle ridges.

Craters come in all sizes, from microscopic pits in the lunar rocks to the giant impact basins such as Mare Imbrium. Some of the larger craters, such as Ptolemaeus have flat interiors and are called walled plains. Many craters

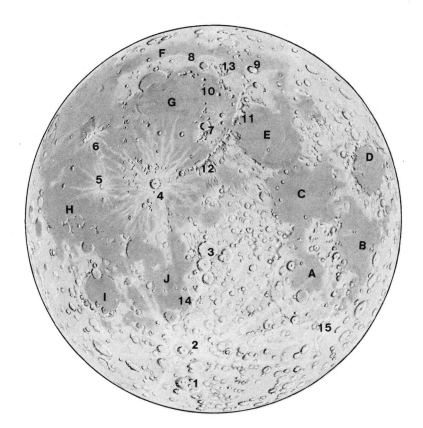

Maria
- A. Mare Nectaris
- B. Mare Foecunditatis
- C. Mare Tranquillitatis
- D. Mare Crisium
- E. Mare Serenitatis
- F. Mare Frigoris
- G. Mare Imbrium
- H. Oceanus Procellarum
- I. Mare Humorum
- J. Mare Nubium

Craters
1. Clavius
2. Tycho
3. Ptolemaeus
4. Copernicus
5. Kepler
6. Aristarchus
7. Archimedes
8. Plato
9. Aristoteles

Other Features
10. Alps
11. Caucasus
12. Apennines
13. Alpine Valley
14. Straight Wall
15. Rheita Valley

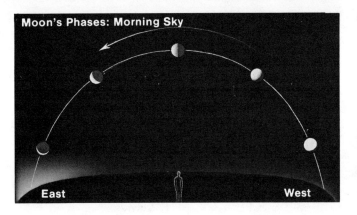

Moon's Phases: Morning Sky

East   West

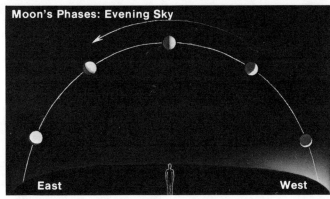

Moon's Phases: Evening Sky

East   West

have central peaks, caused by rebound after the initial impact. Some, such as Copernicus, have complex structure in their walls. Mare Imbrium is an example of a multiple ringed basin. The inner rings have been flooded by lava, but individual peaks can be seen protruding.

When the moon is full there are no shadows, so the brightness of a feature, rather than its profile, determines its visibility. The most prominent bright features are ray systems. The rays are like splash patterns from the impacts that formed the craters. The most conspicuous crater when the moon is full is Tycho, a moderately small crater whose rays cross virtually the entire visible half of the moon. Micrometeorites constantly bombard the moon's airless surface and slowly erase the rays of older craters. Since a crater must be newer than any of the features overlaid by its rays, Tycho must be one of the newest craters on the moon.

Planets and Asteroids

Venus, the morning and evening star, is the brightest object in the sky apart from the sun and moon. In brilliance it

Venus as a 'Morning Star'

Mercury as an 'Evening Star'

East

West

rivals an airplane's landing lights and is often reported as a UFO. It's orbit around the sun lies within our own, so it is always found in the general direction of the sun. It lies toward the east in the morning sky or toward the west in the evening sky. The only details that can be detected in good binoculars are its changing size and phase. During the months it is in the evening sky it is approaching the earth, growing larger and evolving into a crescent. It then passes between the earth and sun and moves to the morning sky, where it is at first a large crescent and gradually shrinks and changes into its full phase as it circles to the far side of the sun.

Jupiter is the most interesting planet in binoculars because of its four bright moons, which rearrange themselves from one night to the next. They are called the Galilean moons, named after their discoverer, Galileo.

Saturn is not magnified quite enough in most binoculars to make its rings visible. (That takes at least 20 or 25 power.) Galileo did not detect them either, although he did notice some puzzling extensions to the disk of the planet. The most you can hope to see of Saturn's shape in binoculars is a slightly elongated yellowish blob. If you look near Saturn you will see its moon Titan. Watch Titan move around Saturn over several nights to verify its status as a

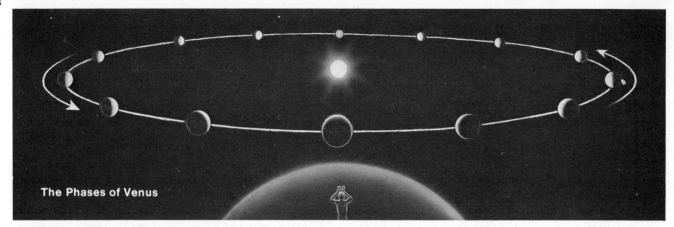

The Phases of Venus

moon. Titan, by the way, is the only moon in the solar system known to have an atmosphere.

Mars appears in binoculars as a tiny redish-orange dot. Mercury, when it is visible, is usually near the horizon masked by twilight. Binoculars are useful in locating it, but once found it appears as a featureless speck. The same comment applies to Uranus, Neptune, and the brightest asteroids. The most interesting aspect of these observations is to follow the motions of the objects from week to week. The word planet means 'wanderer' and the looping motions of the planets against the starry background makes the reason for the name clear. Finder charts for the faint planets and the brightest asteroids are given each January in *Sky and Telescope Magazine,* and updates on all observ-able solar system objects are given there each month.

Comets

Comets are dusty chunks of ice that inhabit the outer reaches of the solar system, mostly far beyond Pluto. As long as they stay far from the sun they remain just as solid as the rocky asteroids closer in. But if a comet comes close to the sun, its outer layers vaporize to form a thin atmosphere called a coma, which is held very loosely by the comet's weak gravity. When a comet is first discovered it usually appears as a faint, fuzzy ball with a point of light at the center. If enough material is vaporized and if the comet passes close enough to the sun, the coma extends to form a tail. The dust is pushed outward by light pressure. The gas

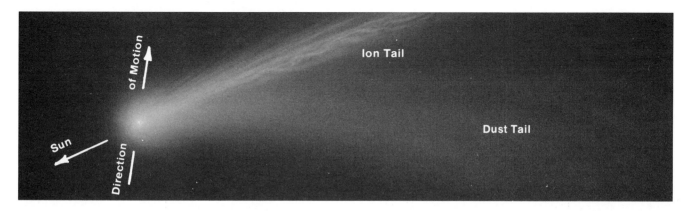

is ionized by sunlight and carried away by an outward flow of charged particles from the sun called the solar wind. The dust and ion tails sometimes separate, the dust tail appearing smooth and gently curved, the ion tail straight but sometimes displaying interesting knots and streaks. When a comet is near the earth watch for its slow motion among the stars, which often becomes apparent in binoculars after a few minutes. Also watch for variations in the details of the coma and tail, which can occur over fairly short time periods.

A comet's brightness and the extent of its activity are hard to predict. The public is easily excited by the prospect of a bright comet, but easily disillusioned if its performance is less than spectacular. Most comets are faint, so find dark skies. Any comet, even a bright one, will be much improved in binoculars. Frequently a comet described as visible to the naked eye will not be easily found unless it is first sighted in binoculars.

One problem in observing comets is finding out where to look. In general, television news coverage is not the best. T.V. newscasters operate under time pressure and often give inadequate or misleading information. Large metropolitan newspapers are a better source and will frequently include a chart. A public planetarium is often another reliable resource. If you are interested in being notified of comets when they are discovered, complete with finder charts and other useful information, you can subscribe to Comet News Service. The address is given in the Resource Guide in Appendix C.

The Milky Way

One of the most impressive sights in all the sky is the panorama of the Milky Way under clear, dark skies. It is a sad commentary on our civilization that a large fraction of our population grows up without ever seeing this wonder of nature.

The Milky Way is our own galaxy as seen from within. We can look at other galaxies from a distance, and from them we get a clue as to what our galaxy would look like from elsewhere in space. But while we strain to see hints of detail in other galaxies with the largest telescopes, we are immersed in a sea of vivid detail here in our own, accessible with even the simplest optical aid.

If we could go outside our galaxy and look down on it from above, it would appear as a giant pinwheel. We would see a bright pinpoint nucleus surrounded by a bulging concentration of yellowish to reddish stars. Extending outward from the central bulge would be a huge flattened disk with stars of all types, highlighted with spiraling chains of gas and dust laced with blue-hot stars. If we looked at our galaxy edge on, rather than from above, the disk would appear as a long, thin spindle enlarged in the middle. Extending the length of the spindle we would see a conspicuous dark, lane of dust.

The spiral arms are filled with knots of dust and glowing gas, often associated with clusters of up to several hundred young stars. According to one theory the spiral arms are like waves that ripple through the disk, compressing the interstellar gas and dust as they sweep by, triggering the formation of new stars, and passing on leaving the newly formed stars in their wake.

As we dive down into the galaxy, from our imaginary vantage point in space, we find our sun with its tiny retinue of planets situated near the inner edge of a spiral arm, called the Orion Arm, about half way to the edge of the disk and a little below its central plane. Except for a nearby spiral galaxy in the constellation Andromeda, and two nearby dwarf galaxies visible in the far south known as the Clouds of Magellan, all we can see from earth with the naked eye belongs to our home galaxy. The stars in our sky are our immediate neighbors. Our depth perception fails us at distances beyond a few hundred yards, so we see the stars as though they were flattened onto a spherical shell surrounding the earth. The constellations are merely patterns of stars as they appear on this imaginary shell with no regard to their actual distances or groupings in a three dimensional sense. The disk of the galaxy projects onto the shell as an encircling band of light from background stars too faint to be seen individually.

Dust lanes block our view of the galactic nucleus, which lies in the direction of the constellation Sagittarius, but we know of its existence by its strong radio emissions. We can see part of the central bulge peeking around the dust lanes forming a large, bright star cloud in Sagittarius. Other star

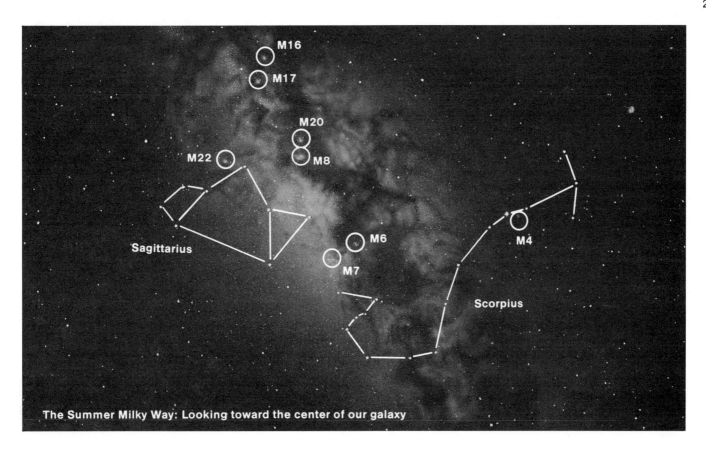

The Summer Milky Way: Looking toward the center of our galaxy

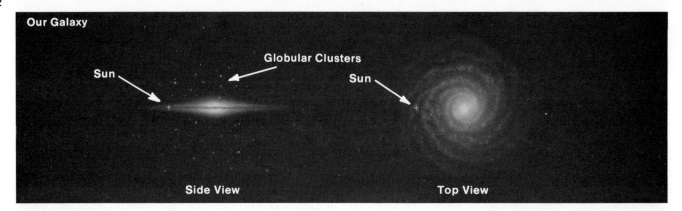

Our Galaxy

Sun

Globular Clusters

Sun

Side View

Top View

clouds from Cygnus through Saggitarius and on to Carina below our southern horizon stand out from the general background glow of the Milky Way. At these points we get lengthwise views along various spiral arms.

The milky path of light continues, though considerably fainter, around the other half of the sky. Looking directly away from the center of the galaxy, the hot blue stars of Orion and Perseus and the star clusters and glowing gas clouds throughout the region are the signposts of more spiral arms.

Binoculars resolve some of the stars of the Milky Way, but they also reveal background glow from stars still deeper in space. The star clouds in Sagittarius, Scutum, and Cygnus are remarkable sights in the wide angle field provided by binoculars.

Woven throughout the Milky Way are dark dust lanes. Some are dense and compact, while others are wispy. The most prominent dust lane is the one that appears to split the Milky Way into two branches in Cygnus. The southern hemisphere's most famous dust cloud is the Coal Sack, a large dark patch near the Southern Cross. Some southern hemisphere tribes have actually patterned the dust clouds of the Milky Way into dark constellations! The dark areas in the Milky Way were once thought to be gaps among the stars, but even casual inspection with binoculars leaves the clear impression that they are foreground clouds obscuring the more distant stars. If we were to travel through one of these dust clouds we would find, at most, a few tiny dust specks per cubic mile! The fact that they appear opaque testifies to their tremendous total volume.

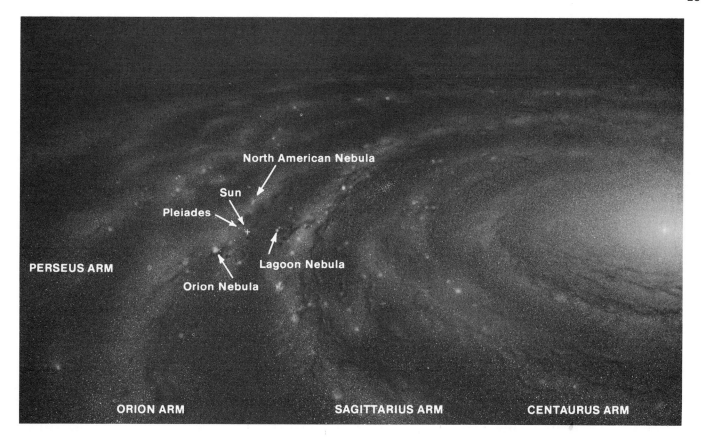

North American Nebula

Sun

Pleiades

PERSEUS ARM

Orion Nebula

Lagoon Nebula

ORION ARM

SAGITTARIUS ARM

CENTAURUS ARM

Nebulae

The word nebula means cloud. It was originally applied to any object that looked fuzzy in a small telescope, including what we know today as galaxies. Today the word is reserved for actual clouds of gas or dust between the stars.

Space is not quite empty. Between the stars there is an average of one atom per cubic centimeter (16 atoms per cubic inch). In some regions of space this material is bunched up in clouds about 1000 times the average density, but this is still a billion times thinner than even the best vacuum chambers on earth. On the other hand, space is so vast that the total amount of matter between the stars is very great, rivaling the amount of matter in the stars themselves.

The interstellar clouds consist mostly of hydrogen atoms with small amounts of other atoms and occasional microscopic grains of dust mixed in. The gas is transparent. It becomes visible only near very hot, blue stars that give off large amounts of ultraviolet light, making it glow much as 'black light' causes certain poster inks to glow. These glowing gas clouds are known as 'emission nebulae'. Another kind of bright nebula results from dust reflecting the light of a nearby star. In color photographs emission nebulae are usually red or green while reflection nebulae are usually blue. The colors are faint, however, and not

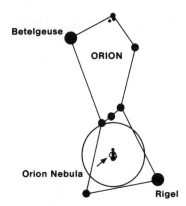

generally seen by direct observation.

A hot, blue star has such a short lifespan that it never moves far from its birthplace. Thus the cloud it illuminates must be its parent cloud. When large glowing clouds are investigated more thoroughly they frequently turn out to be active star formation sites.

The Orion Nebula, M42

The Orion Nebula is the fuzzy 'star' in the middle of Orion's sword. It is easily seen to be a glowing cloud in binoculars or telescopes of any size. The nebula is known to be the nursery for a whole cluster of newly forming stars. The visible nebula is just a small illuminated portion of a much larger cloud that extends beyond the borders of the constellation Orion. The whole Orion complex is located

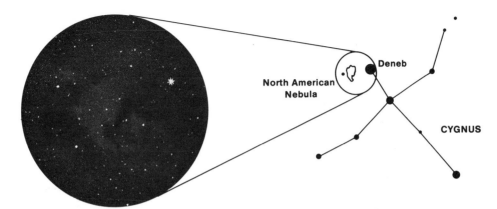

within our own spiral arm of the galaxy.

The designation M42 is the Orion Nebula's catalog number in the Messier Catalog. The renowned 18th century comet hunter Charles Messier cataloged over 100 of the brighter nebulae, galaxies and star clusters to avoid confusing them with comets.

Nebulosity in and near Sagittarius

The Lagoon (M8), the Trifid (M20), the Swan or Omega (M17), and the Eagle (M16) are emission nebulae in a neighboring spiral arm of our galaxy called the Sagittarius Arm. The Lagoon and the Eagle nebulae are conspicuously intertwined with star clusters. The names refer to patterns seen in telescopes. (See Milky Way map on page 21.)

The North American Nebula

Near the star Deneb in the constellation Cygnus is a complex pattern of dark dust clouds that stand out clearly from the background Milky Way, when observed in binoculars. Emerging from behind one of the dust clouds is a very subtle glowing patch shaped remarkably like the continent of North America. The Atlantic coastline and the Gulf of Mexico are sharply outlined by the foreground dust clouds. The North American Nebula is better seen in binoculars than in most telescopes because of its great size and low surface brightness. It requires dark skies and it is probably the most challenging of the nebulae listed here, but it is a rewarding one to find.

Stellar Remnants

The kind of nebula discussed in the previous section typically contains enough material to form hundreds or thousands of stars and is generally the site of active star formation. A planetary nebula, on the other hand, is a cloud of material errupted from a single star as the star runs out of fuel and becomes unstable. All that remains of the central star is its extremely hot, dense core, which is evolving into what is known as a white dwarf. The matter in such a star is compressed so much that a single teaspoonful weighs tons.

Planetary nebulae have nothing to do with planets. The name comes from the fact that some of the round ones resemble planets in a telescope. They come in a variety of shapes. Some look like disks or rings, while others have the double lobed appearance of a dumbbell. Planetary nebulae range in apparent size from almost starlike points to large, diffuse loops. Very few are suitable objects for binoculars.

Sometimes grouped with the planetary nebulae are supernova remnants. These nebulae share the distinction of originating from a single aging star, but a supernova is a very different way for a star to deal with its old age. When stars much more massive than the sun become unstable they are not able to gently rid themselves of their excess baggage and settle peacefully into the role of a white dwarf. Rather, they explode with great violence, ejecting most of

their mass into space. The light from a supernova can, for a short time, rival the total light output of the rest of the galaxy. The remaining core is so compressed by the explosion that it may be turned into a neutron star millions of times denser than a white dwarf.

The Dumbbell Nebula, M27

The Dumbbell Nebula is one of the brightest planetaries, and is described aptly by its name. In binoculars it looks like a small, fuzzy patch, twice as long as it is wide and slightly pinched in the middle. Each lobe appears about six times the size of Jupiter's disk, although the nebula is much fainter than Jupiter.

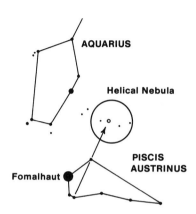

The Helical Nebula

The Helical Nebula is a very challenging object for beginners, but it is visible in good binoculars and easier to see with binoculars than with most telescopes. Many telescope users have never seen it because of its large size and low surface brightness. Its total light is actually greater than that of the Dumbbell, but its light is coming from an area in the sky nearly the size of the moon! In a dark sky it is faintly visible in binoculars as a large round patch. In photographs its looped structure makes it look like a coiled spring, hence its name.

Open Star Clusters

As you sweep the sky with binoculars you will run across numerous places where stars are conspicuously bunched together in clusters of fifty to several hundred stars. These are called open clusters, in contrast to the much more concentrated globular clusters which will be discussed later. We shall list only a few examples that are particularly suitable for binoculars. The real problem is deciding where to stop. Several of the nebulae already mentioned are associated with star clusters that will not be repeated here. You should have no trouble finding many more examples by random searching, especially in and near the Milky Way.

Clusters occur because when the conditions are right for stars to form, they usually form in bunches. Frequently material remaining from the parent cloud is visible as an associated emission or reflection nebula. The stars in a cluster are held together loosely by their combined gravity, but one by one, the stars can break away. As a cluster loses stars it slowly loses its gravitational grip on the remaining members. Usually within a few hundred million years, the time it takes to go once around the galaxy, a cluster will entirely disperse.

The Pleiades

The Pleiades, also known as the Seven Sisters, is the most famous open cluster in the sky. Most people can see

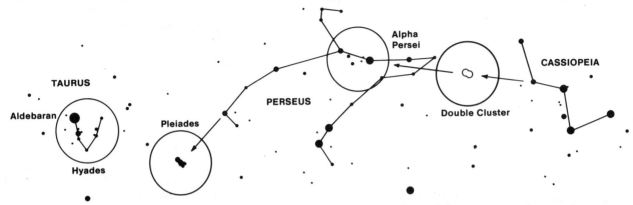

only six stars with the unaided eye, but some with exceptional vision have reported seeing several more. Binoculars reveal many more stars. There may actually be over 300 cluster members.

The Hyades

The Hyades form the 'V' shaped head of Taurus the Bull. Aldebaran, the brightest star in the field, is not a cluster member but lies in the foreground. In binoculars the simple 'V' shape becomes filled in with a rich background of fainter stars. The Hyades is one of the nearest clusters to the earth.

The Alpha Persei Group

At the heart of the constellation Perseus surrounding its brightest star is a large, dense conglomeration of bright stars mixed in with the Milky Way. It is spectacular in binoculars and has in recent years been confirmed to be a true open cluster with about 100 members.

The Double Cluster

Along the Milky Way half way between Cassiopeia and Perseus is a pair of close, dense star concentrations. They are faintly visible to the unaided eye, but they are striking in binoculars. The two clusters are not believed to be connected, but rather lie at different distances along the same line of sight. Both clusters are in the next arm of the galaxy outside our own, known as the Perseus Arm.

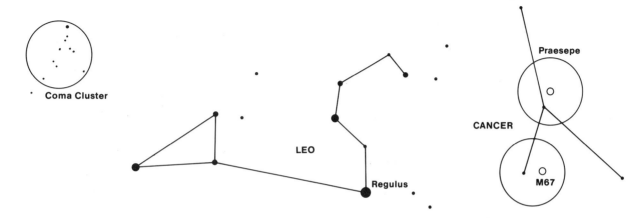

Coma Cluster

Praesepe

CANCER

LEO

Regulus

M67

The Coma Cluster

The constellation Coma Berenices, translated 'Berenice's Hair', refers to Berenice II, queen of Egypt in the third century B.C., who according to legend sacrificed her hair as an offering to the gods. The otherwise inconspicuous constellation takes its name from a generous sprinkling of stars barely visible to the naked eye. These stars are actually an open cluster only slightly farther from us than the Hyades, and containing about 40 members. It is not included in most lists because it is too large to be considered a telescopic object, but it is beautiful in binoculars.

Praesepe, M44

Praesepe, or the Beehive cluster, is visible to the naked eye as a fuzzy patch in the constellation Cancer. It was known in antiquity but discovered to be a cluster of stars for the first time by Galileo. It contains about 200 members and is about twice the apparent size of the moon.

M6 and M7

These two open clusters must compete for attention with the other wonders of the summer Milky Way. They are easy naked eye objects directly above the tail of Scorpius and are excellent in binoculars. M6 is about twice as distant as M7. (See Milky Way map on page 21.)

Globular Star Clusters

Globular clusters look like cotton balls in binoculars, but they are actually clusters of a hundred thousand to a million stars. The stars are packed much closer than the sun and its neighbors, but if you were to picture the stars in the center of the clusters as grains of sand, they would still be scattered a half mile or more apart.

These giant clusters are associated with the Milky Way system, but they are not confined to the galactic plane. Rather, they swarm around the galaxy in a more or less spherical distribution. In the sky the vast majority of globular clusters are found in the general direction of the galactic center in Sagittarius. Historically, this was the first evidence used to show that the sun does not lie at the center of the galaxy. It is easier to think of the sun being off center in the galaxy than to think of all the globular clusters swarming around some off centered point.

It may puzzle you to think of stars belonging to our galaxy lying far outside the galactic plane. There are no raw materials out there from which they could form. The loose gas and dust available for star formation today lies in the flat galactic disk. There is no way huge clusters of stars could be formed within the disk of the galaxy and then be launched out of the disk. The fact that stars are found outside the disk implies that at some time in the distant past our galaxy was not the flat pinwheel it is today. It must have

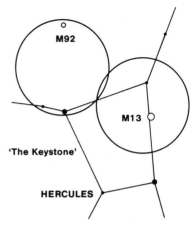

been nearly spherical when the globular clusters formed. The rotation of the galaxy then caused it to gradually flatten out, leaving behind the stars that were already formed. The stars in the globular clusters are thus very old, dating back to a time near the formation of our galaxy. Other studies confirm this is indeed the case.

The Hercules Cluster, M13

The globular cluster in the 'Keystone' of Hercules is the most famous example of this type of object for observers in the northern hemisphere because it rises high in the summer sky. It is one of the half dozen brightest globulars and probably contains over a million stars.

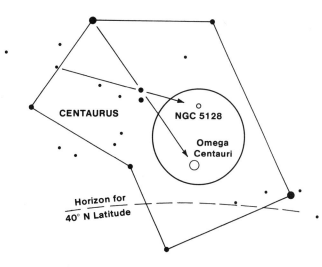

M22

M22 is comparable in brightness to M13, but it is closer to the southern horizon. If you live in the lower half of the United States this is no problem. It contains an estimated half million stars and is one of the nearest globular clusters to the earth. (See Milky Way map on page 21.)

Omega Centauri

Omega Centauri is one of the two most spectacular globular clusters in the sky, but it lies far to the south and skims the horizon for observers in the northern United States. The further south the observer, the better. It lies almost due south of the first magnitude star Spica, so the best time to look for it is when Spica is highest in the sky. The apparent size of the cluster is at least that of the moon, although photographs, bringing out fainter stars, show it to be over twice as large. Besides containing over a million stars it is one of the nearer globulars.

Galaxies

Although it may seem to be just a faint band of light, the Milky Way dominates the sky. Everything we have discussed, from the constellations to the globular clusters, is related to the Milky Way in one way or another. The Milky Way also blocks the view of what lies beyond. To see out of our galaxy we have to avoid the dust clouds that accompany that faint band of light. Those empty looking skies that come overhead on spring and autumn evenings may not look as rich as the skies of summer and winter, but for that very reason these are the prime hunting grounds for galaxies beyond our own.

Quite a few galaxies are visible in binoculars, but the question is, how visible is visible? There are several galaxies that are excellent binocular objects. Many more could be termed 'detectable' in binoculars. They look like faint smudges, but perhaps even a smudge is awe inspiring when you realize it contains upward of a hundred billion stars.

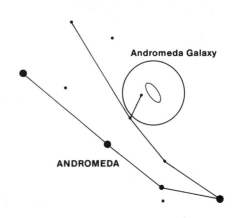

like the Milky Way, but it looks oval because it is inclined to our line of sight. It is bright and condensed in the middle, and its visible disk fades out gradually. See how far out you can detect it using averted vision.

The Andromeda Galaxy, M31

The most inspiring example of a spiral galaxy in binoculars is our closest full-sized neighbor in the constellation Andromeda. It is better seen in binoculars than in most telescopes because of its apparent size, which is about five times the apparent diameter of the moon. In actual size it is slightly larger than the Milky Way, which is itself a larger than average spiral galaxy. The Andromeda galaxy is disk shaped,

The Triangulum Galaxy, M33

Only slightly farther than M31 is M33, a spiral galaxy in the constellation Triangulum. M33 is positioned so we see the full face of its disk. The galaxy consists almost entirely of spiral arms, with hardly any central condensation. It has less than a tenth as many stars as the Milky Way, so its overall brightness is low. Yet it appears fairly large in the sky because it is so close. Its low brightness combined with

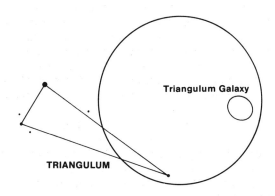

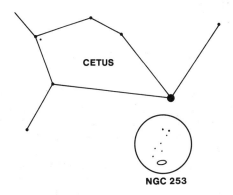

its large apparent size, which is greater than that of the moon, makes it a difficult object to see in small telescopes. It is readily spotted in binoculars, however, on dark nights.

NGC 253

The designation NGC stands for 'New General Catalog', a much more complete listing of clusters, nebulae, and galaxies than the older Messier catalog. NGC 253 is a large spiral galaxy that is inclined to our line of sight about the same amount as the Andromeda Galaxy, giving it a similar appearance. It is almost four times as far away as the Andromeda Galaxy and not quite a fourth as large in apparent size. It has higher surface brightness than M33 and is easy to see with either binoculars or a telescope.

NGC 5128, Centaurus A

NGC 5128 (known to radio astronomers as Centaurus A) lies about four degrees due north of the globular cluster Omega Centauri, so it is another horizon skimmer for those observing from high latitudes. In the southern U.S. it is not too difficult to spot, being one of the brightest galaxies and about a third the size of the moon. In general appearance it is round, typical of giant elliptical galaxies. Unlike spiral galaxies, elliptical galaxies are not flattened into a disk and generally have little or no dust. NGC 5128 is unusual for an elliptical galaxy in that it does have a wide, dark belt of dust across it. The dust belt may be difficult to pick out with binoculars. (See map on page 31.)

Appendix A

Seasonal Tours

We highly recommend that you obtain a copy of *The Night Sky* star dial for your own latitude zone as your primary reference for learning the constellations. The maps and descriptions on the next few pages are not intended to replace such a chart. They are highly simplified, intended only to help you get your bearings. The charts are drawn for about 9:00 p.m. in the middle of the season, but they should prove useful in the evening hours throughout the season, if you realize that the overhead stars shift westward with time. Keep in mind that planets may be in the sky. Any really bright stars besides those shown on the charts are probably planets.

Summer

First find north. The Big Dipper will be located a little to the left of north with its handle pointing upward. It will be high in June and closer to the horizon in August.

The two stars at the end of the Big Dipper's cup point to Polaris, a star about as bright as the Big Dipper stars and as far from the cup of the dipper as the dipper is long. Three stars from the little dipper are shown but not labeled: Polaris lies at the end of the handle and the two stars slightly up and to the left from Polaris form the far end of the cup. The stars in between are very faint and cannot be seen unless the sky is quite dark. Once you locate Polaris you can always count on finding it there. The rest of the stars circle around the sky, but being at the pole, Polaris stays fixed.

Look across the pole from the Big Dipper where the 'W' shaped constellation Cassiopeia is rising.

Return to the Big Dipper and follow the arc of the handle first to Arcturus and then on to Spica. Both are first magnitude stars. When you have found Spica you should be facing southwest.

Face northeast and look fairly high overhead. Three of the brightest stars in the sky form a large south-pointing triangle, often called the Summer Triangle or the Mariner's Triangle. The brightest of the three stars is Vega, the southernmost is Altair, and the easternmost and faintest of the three is Deneb. If you are in dark skies you will see the Milky Way approximately in line with Deneb and Altair.

Now face south. The most conspicuous star in this part of the sky (unless a planet is up) is Antares. Antares is a distinctly red star in the constellation Scorpius, which lies to the right of the brightest part of the Milky Way. The center of our galaxy lies slightly above the tail of the scorpion and just to the right of the teapot-shaped figure, Sagittarius.

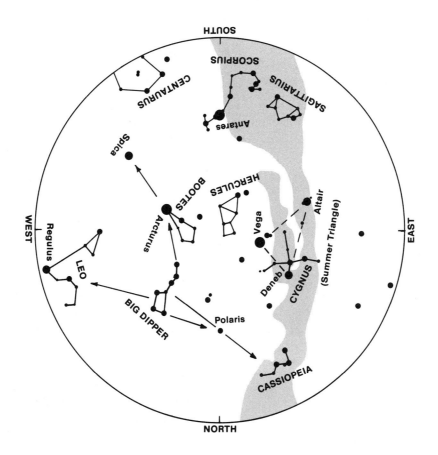

Autumn

Face northwest and look high in the sky to find three very bright stars forming a large south-pointing triangle. This is the summer Triangle, consisting of Vega, Altair, and Deneb as described above in the summer tour.

Face south to find the first magnitude star Fomalhaut, the only bright star in the entire region.

Look to the northeast where the stars Capella and Aldebaran are starting to rise. Slightly above Aldebaran is the conspicuous cluster of stars called the Pleiades. This cluster is frequently mistaken for the Little Dipper.

Face a little to the east of north and look fairly high in the sky along the Milky Way to find the 'W' shaped constellation Cassiopeia. The top of the 'W' faces down and to the left toward Polaris.

Follow the Milky Way toward the east from Cassiopeia to find Perseus, with its lower arm curved toward the Pleiades.

The Great Square of Pegasus is almost overhead. It's stars are only moderately bright, but the other stars in this part of the sky are even less so.

Follow the double chain of Andromeda to the northeast from Pegasus. Don't neglect Andromeda's two small tag-along constellations, Triangulum and Aries.

Most of the constellations south of Pegasus are rather faint. To locate them, refer to the back side of The Night Sky star dial and use Fomalhaut as a reference point.

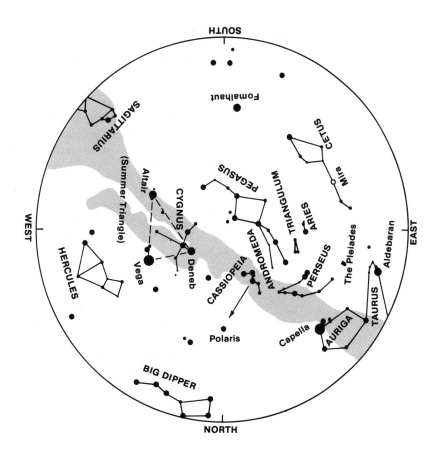

Winter

The obvious starting point in the winter is the bright constellation Orion to the south. The famous figure of the hunter is highlighted by the three bright stars forming his belt and the trail of fainter stars hanging down from it forming his sword. These are surrounded by four bright stars including the first magnitude stars Rigel and Betelgeuse.

Starting with Rigel trace out a remarkable loop of first magnitude stars as follows: from Rigel, up and to the right to Aldebaran, on up and slightly left to Capella, down and to the left to the bright pair Castor and Pollux, down to Procyon and finally down to Sirius, the brightest star in the sky. Note for reference that of the pair, Castor and Pollux, Castor is toward Capella and Pollux is toward Procyon.

If you are in Florida, Hawaii, or southern Texas you will see another first magnitude star, Canopus, near the horizon due south of Sirius, and in early winter another, Achernar westward from Canopus.

Tear yourself away from the brilliant spectacle in the south long enough to notice the Big Dipper rising in the northeast and Leo, with its first magnitude star Regulus rising in the east.

Cassiopeia, Perseus, and Andromeda are still visible in the northwestern sky.

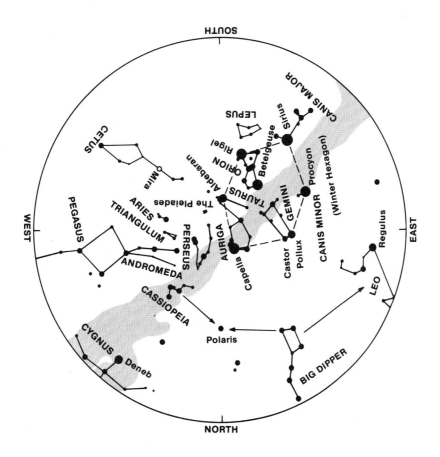

Spring

The northern half of the spring sky is dominated by the Big Dipper high over the pole. This is the best time of year to trace out the fainter extensions of the full constellation, Ursa Major. Note in particular the alignment of the three pairs of stars pictured as the feet of the beast.

Look directly south of Ursa Major to find Leo, with its prominent sickle shape terminating with the first magnitude star, Regulus.

Follow the arc of the handle of the Big Dipper to Arcturus and on to Spica, which lies in the large, ungainly constellation of Virgo.

To the southwest of Virgo you should easily locate the much more compact constellation Corvus, but the remaining constellations in this part of the sky are more difficult to pick out.

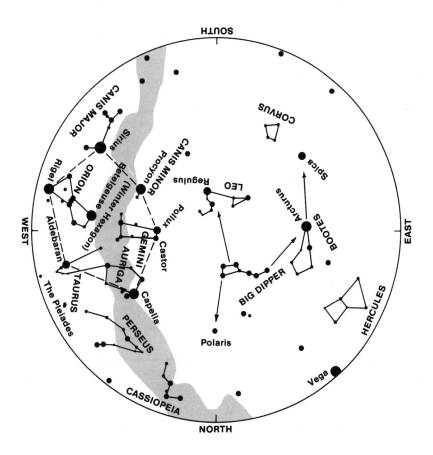

Appendix B

Buying Your First Telescope

Probably the majority of people who buy telescopes do so with eager anticipation, yet turn around and allow them to sit unused because they don't know how to find objects of interest in the sky. If you have pushed your binoculars to the limit and are still hungry for more, the prognosis for your success with a telescope is much better. You have become aquainted with the sky in general, you have located a handful of fascinating deep sky objects that cry out for closer examination, and you have developed a taste for staying out under the stars at night.

It is important to be wise in the purchase of your first telescope to avoid disillusionment. The following pointers come from years of experience watching newcomers take up astronomy as a hobby.

1. Find an active astronomy club before you buy a telescope. If you don't know where one meets, inquire at a public planetarium or the astronomy or physics department of a local college. Attend a 'star party' (be sure to take your binoculars along). Be a listener. Draw on whatever experience they have to offer. See the kind of view offered in various telescopes. Is the optical quality good? Test this by putting a star slightly out of focus. An out of focus star should look like a set of nested rings. Are the rings round or badly distorted? Are star images sharp at high power? How does the telescope perform mechanically? The mounting is as important as the optics. Is it well balanced? Does it vibrate badly at the slightest breeze? Does it settle down in a few seconds when bumped slightly? Does it move smoothly, allow easy pointing, and hold its position? Is it easily transportable? Is the owner satisfied?

2. Stay away from 'department store' telescopes. Such telescopes typically gather too little light and have mountings that may look good but perform very poorly. A better place to shop is in the pages of *Sky and Telescope* or *Astronomy Magazine*. Don't overlook the classified ads for second hand telescopes. It is amazing how many people have virtually unused telescopes for sale!

3. For a telescope that can begin to give satisfactory views of galaxies and nebulae, look for one at least 6 or 8 inches in diameter. A 6 inch telescope will give good views of the brighter objects and a limited amount of detail in fainter ones. It is a good size for a first telescope. It will take you a couple of years, at least, to push it to its limit. Whatever size telescope you get, there will always be marginally visible objects and hints of faint detail that make you lust for more light. A 6 inch telescope used to be considered a big telescope among amateurs. Such is not the case any more. 10 and 12 inch telescopes are common in many clubs, and telescopes 16 inches and up are seen more and more frequently. You should definitely be aware of the Dobsonian telescope design. John Dobson, founder

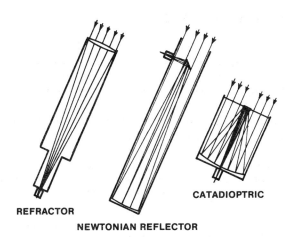

REFRACTOR

NEWTONIAN REFLECTOR

CATADIOPTRIC

with. A refractor has a lens at the front and you look in at the back. Binoculars are short focal length refractors, and what follows does not apply to them. Astronomical refractors usually have very long focal lengths, giving high magnification and low field brightness. They are very expensive in the larger sizes. You should probably not buy a long focal length refractor for general purpose viewing.

The most common, simplest, and cheapest design is the Newtonian reflector. The Newtonian has a curved mirror at the back and a diagonal mirror at the front. You look into the side of the tube near the front end. Newtonians can be somewhat bulky, but unless this is too much of a hardship, a Newtonian is probably your best buy. (You can find ways to transport a big telescope with a small car!)

A third category that has become popular in recent years utilizes folded optics. The tube is short and squat, there is a curved primary mirror in the back, a glass plate, either curved or flat, across the front, and you view from the back through a hole in the primary mirror. This is called a Catadioptric design, with Maksutov and Schmidt-Cassigrain being the main sub-categories. They are more compact, cost a lot more, and have both advantages and disadvantages optically. They typically have longer focal lengths (thus higher magnification) than comparable Newtonians, even though the tube is shorter. If cost is not a problem but storage space is, a catadioptric is the way to go. A well made telescope in any of these catagories will give beautiful images.

of the San Francisco Sidewalk Astronomers, popularized large, very economical telescopes with thin mirrors and extremely simple and cheap yet stable mountings. You can probably buy (or make) a 10 inch or larger Dobsonian telescope for what you would pay for a traditionally mounted 6 inch one. You should be aware that Dobsonian telescopes are not designated for photography, but your first telescope should be for looking, anyway. Astrophotography can come later.

4. There are several optical designs you will be confronted

Appendix C

Resource Guide

(Prices are given as a guideline only, and are believed to be current as of 1983.)

STAR MAPS AND ATLASES

The Night Sky, by David Chandler
A heavy cardboard star dial specially designed for low distortion (see illustration on title page) / Available for different latitude zones (specify your latitude or nearest major city) / Includes a number of objects visible with binoculars (price: $5.95 ppd. for one chart, $3.25 ea. for additional charts)
David Chandler, P.O. Box 309, La Verne, CA 91750

Scanning the Sky
A catalog of atlases and astronomical reference books available from Sky Publishing Corp. / A must for anyone interested in observing the sky (price: Free)
Sky Publishing Corp., 49 Bay State Rd., Cambridge, MA 02238

ASTRONOMICAL CALENDARS

Astronomical Calendar, by Guy Ottewell
A guide to the year's celestial events / Large atlas format with many drawings and diagrams / Packed with useful information for observers and helpful explanations for beginners (price: $10)
Astronomical Calendar, Dept. of Physics, Furman University, Greenville, South Carolina 29613

The View from the Earth, by Guy Ottewell
A smaller format, simplified version of the Astronomical Calendar designed for children or other beginners (price $5)
Astronomical Calendar, Dept. of Physics, Furman University, Greenville, South Carolina 29613

Sky Calendar
A set of monthly sheets with a simplified star chart on one side and a calendar showing planet and moon positions on the other / great for posting in a classroom (price: $2 per year)
Sky Calendar, Abrams Planetarium, Michigan State University, East Lansing, MI 48824

GENERAL REFERENCE

Astronomical Companion, by Guy Ottewell
A thorough introduction to basic astronomy / Highly visual in approach, with many line drawings, charts, etc. / Same format as Astronomical Calendar, but specializing in information that does not go out of date each year (price: $12)
Astronomical Calendar, Dept. of Physics, Furman University, Greenville, South Carolina 29613

Burnham's Celestial Handbook, by Robert Burnham Jr. (3 Volumes)
The ultimate descriptive guide to the heavens for amateurs / over 2000 pages with photographs, descriptions, data, and articles (price: approx $10 per volume at bookstores)

Astronomy: The Cosmic Journey, by William K. Hartman
One of the better non-mathematical astronomy texts used at the introductory college level / Published by Wadsworth Publishing Co., Belmont, CA / Can be ordered through most bookstores

PERIODICALS

Sky and Telescope Magazine, 49 Bay State Rd., Cambridge, MA 02238 (price: $18 per year)

Astronomy Magazine, P.O. Box 92788, Milwaukee, WI 53202 (price: $21 per year)
(*Odyssey*, an astronomy periodical for children, and *Telescope Making* are two very good subsidiary publications of *Astronomy Magazine*.)

OTHER SOURCES OF INFORMATION

Comet News Service
For rapid notification of discoveries of new comets / Includes tracking charts, orbit diagrams, and other useful information / Quarterly review issues discuss the latest developments in cometary astronomy (subscription rate: $4 per year)
Comet News Service, McDonnell Planetarium, St. Louis, MO 63110

Hansen Planetarium
Probably the largest single source of astronomical slides, posters, etc. / Free catalog available
Hansen Planetarium, 15 South State St., Salt Lake City, UT 84111

OTHER PUBLICATIONS OF THE AUTHOR
Deep Space 3 - D, a Stereo Atlas of the Stars / A set of cards showing the sky in 3 - D / stereo viewer included (price: $9.95)
Halley's Comet Model, a 3 dimensional cardboard model illustrating the orbit of Comet Halley (price: $3 for one, $1.25 for additional copies)
Slides of the space artwork of Don Davis (flier sent by request)